WILD WILD WEST

Wildlife Habitats of Western North America

Words and Illustrations by
Constance Perenyi

*This book is dedicated to the preservation of North America's
largest remaining wildlife habitat, the Arctic National Wildlife Refuge.*

Printed in Hong Kong.

Designed by Eilisha Dermont.

Library of Congress Cataloging in Publication Data
Perenyi, Constance. 1954-
Wild Wild West :wild life habitats of western North America / by Constance Perenyi.
p. cm.
Summary: Explores eleven wildlife habitats in western North America, from the
arctic tundra to arid desert, and examines the life they support.
ISBN 0-912365-82-X
ISBN 0-912365-90-0 pbk.
1. Natural history–West (U.S.)–Juvenile literature. 2. Natural history–Canada,
Western–Juvenile literature. 3. Habitat (Ecology –West (U.S.)–Juvenile literature.
4. Habitat (Ecology)– Canada, Western–Juvenile literature.
[1. Natural history –West (U.S.) 2. Habitat (Ecology)–West (U.S.)] I. Title.
QH104.5.W4P47 1993
508.78–dc20

92-46995
CIP
AC

Sasquatch Books
1931 Second Avenue
Seattle, Washington 98101
(206) 441-6202

Each page in
Wild Wild West is
a collage, created with
layered papers. Every
piece was individually cut
or torn from papers such as
Mexican bark, Japanese
tie-dyed origami, or tissue,
and then glued to the
background.

On the planet Earth, there is a special place for every living creature.

For people, that special place is where we live, work, play, and go to school. It includes our neighborhoods, our towns, and our communities. It is everything that makes a place home.

For plants and animals, these special places are called habitats.
Not all animals build shelters like people do, or even stay in
one place for very long. In fact, a habitat may be as small as a
tiny crack in a garden wall, where insects can live for generations,
or as big as the vast Arctic tundra, where caribou range over
many thousands of miles, searching for food. In their habitats,
animals find everything they need to survive: food, water,
and a safe place to raise their young.

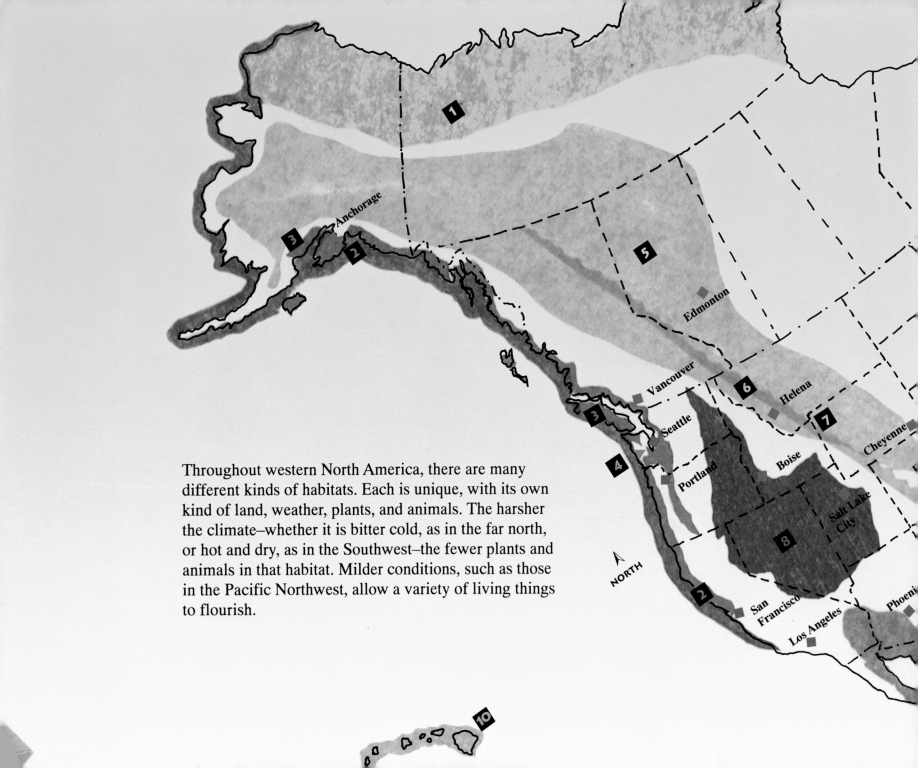

Throughout western North America, there are many
different kinds of habitats. Each is unique, with its own
kind of land, weather, plants, and animals. The harsher
the climate–whether it is bitter cold, as in the far north,
or hot and dry, as in the Southwest–the fewer plants and
animals in that habitat. Milder conditions, such as those
in the Pacific Northwest, allow a variety of living things
to flourish.

KEY

1 **ARCTIC TUNDRA** Northern Canada and Alaska

2 **ROCKY COAST** Pacific coastline from Alaska to California

3 **ANCIENT FORESTS** Alaska, British Columbia, Washington, Oregon, California

4 **BOWERMAN BASIN** Grays Harbor, Washington

5 **BEAVER PONDS** Alaska, British Columbia, Montana, Idaho, Utah, Wyoming, Colorado

6 **ROCKY MOUNTAINS** British Columbia, Alberta, Montana, Wyoming, Colorado, New Mexico

7 **YELLOWSTONE NATIONAL PARK** Wyoming, Montana, Idaho

8 **INTERMOUNTAIN GRASSLANDS** Washington, Oregon, Idaho, Nevada, Utah

9 **SONORAN DESERT** California, Arizona, and Baja and Sonora in Mexico

10 **HAWAIIAN CORAL REEF** Hawaiian Islands

11 **CITY HABITATS** Throughout the West (indicated on the map by ▪)

1 ARCTIC TUNDRA

The Arctic **tundra** is a land so cold that the ground is always frozen, and in winter, the sun never shines. Chilling winds blow across the flat landscape, sweeping snow from the ground and uncovering patches of moss, lichen, and dead grass for hungry caribou to eat. The caribou herds do not stay in one place for long, but keep moving to find food.

Then, for two months each year, summer warms the tundra. The ice melts and young polar bears emerge from caves with their mothers. Caribou pause to eat plants and to care for newborn calves. But winter returns quickly. The long trek over snow and ice begins again.

Animals pictured: caribou.

Waves pound the shoreline of the Pacific Ocean from Alaska to California, filling the air with the salty smell of the sea. The coastline is carved by these waves, by the action of water and wind known as **erosion**.

Chunks of rock break away from steep cliffs, leaving ledges for seabirds to nest upon and places for seals and sea lions to rest. Otters wrap themselves in kelp to stay in place in the water. Each time the tide rises, the ocean brings fresh food for barnacles and other sea creatures. Low tide exposes mussels and crabs that are then eaten by seabirds.

Animals pictured (left to right): double-crested cormorants, Steller sea lion, orcas, sea otter, tufted puffin. Plants pictured: kelp.

3 ANCIENT FORESTS

Near the shores of the Pacific Ocean, from southeast Alaska to northern California, giant trees grow in dark, damp forests. Fog from the ocean rolls across the treetops, clouds hang heavy in the sky, and on most days, rain falls. Some forests receive 15 feet of rain in a year, making them the wettest spots and the only **rain forests** in North America.

Old-growth trees, which range in age from 175 to over 1,000 years old, are draped with mosses and lichens. Animals use dead trees that are still standing for perches and nesting holes, while fallen, rotting trees–called nurse logs–support seedlings, creating new life in the forest.

Animals pictured (left to right): Pacific giant salamander, deer mouse, northern spotted owl. Plants pictured (left to right): sword fern, evergreen huckleberry, salal, Douglas fir trees in the background.

4 BOWERMAN BASIN

For a short time each spring, thousands of shorebirds descend on the **tidal mud flats** at the mouth of Washington's Hoquiam River. There, the changing tide exposes mud that is rich in plant and animal life.

Bowerman Basin is an important temporary habitat for shorebirds. As they travel north to nesting grounds in the Arctic, they stop here to eat their fill of insect larvae and worms that lie hidden in the mud. The birds find enough food to last the final 2,500 miles of their long journey.

Animals pictured: (on the ground) plovers, sandpipers, dunlins, dowitchers, sanderlings;
(in the sky) a mew gull.
Plants pictured: beach grass.

5 BEAVER PONDS

Beaver ponds are found throughout the West–anywhere there are water and trees, and of course, beavers. When beavers build a dam across a stream, they stop the flow of water and create a shallow lake, which makes a new habitat. The flooded land comes alive with plants and animals that cannot live on a moving stream, but flourish in **wetlands**, which are areas that stay wet most of the time.

Summer is an especially active season on the beaver pond. Water lilies bloom, providing shelter for fish, frogs, and insects. Moose, which live in some areas of the West, often visit beaver ponds, wading in to feast on the water plants and to escape biting flies. And beavers work nonstop to keep the dam in good condition.

Animals pictured (left to right): beavers, dragon flies, northern leopard frog, moose.
Plants pictured: water lilies.

6 ROCKY MOUNTAINS

The Rockies form the longest range of mountains in North America, stretching from Canada in the north through New Mexico in the south. Wind, sun, altitude, and extreme temperatures make life on top of the mountains very difficult. The upper elevations of these mountains, capped with snow all year, are an **alpine** habitat. Here, there are rocks and a few wildflowers, but no trees.

The mountain lion, also known as the cougar, puma, or catamount, hunts alone on these rugged peaks, searching for prey as large as bighorn sheep or as small as ground squirrels. A shy and wary animal, this cat avoids humans and disappears from sight quickly, leaving nothing behind but footprints in the snow.

Animals pictured: mountain lion.
Plants pictured: bitterroot.

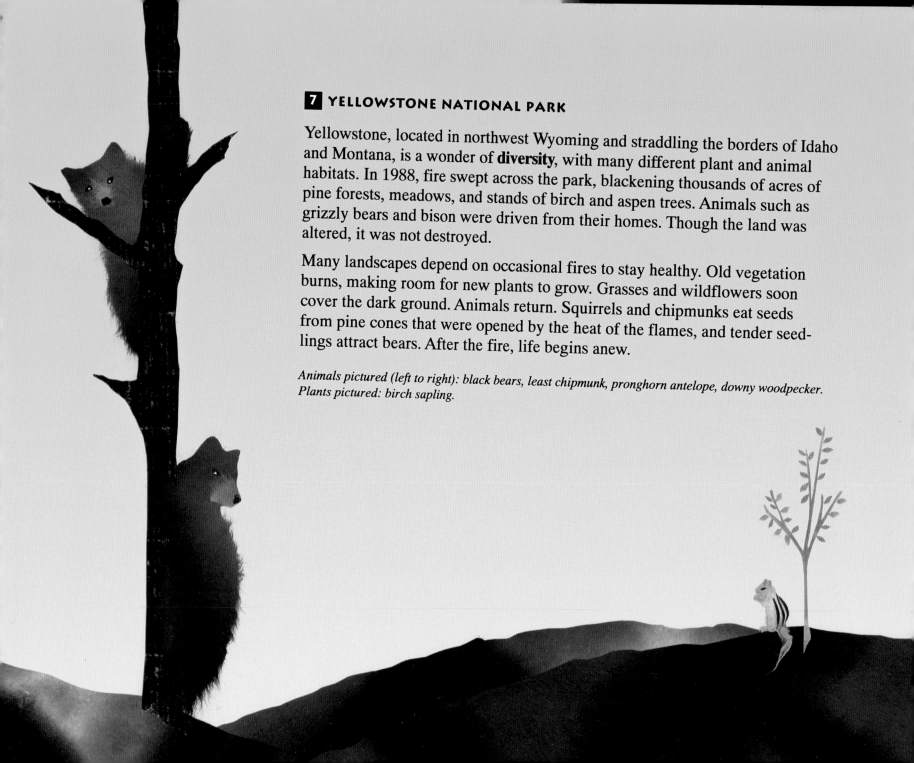

7 YELLOWSTONE NATIONAL PARK

Yellowstone, located in northwest Wyoming and straddling the borders of Idaho and Montana, is a wonder of **diversity**, with many different plant and animal habitats. In 1988, fire swept across the park, blackening thousands of acres of pine forests, meadows, and stands of birch and aspen trees. Animals such as grizzly bears and bison were driven from their homes. Though the land was altered, it was not destroyed.

Many landscapes depend on occasional fires to stay healthy. Old vegetation burns, making room for new plants to grow. Grasses and wildflowers soon cover the dark ground. Animals return. Squirrels and chipmunks eat seeds from pine cones that were opened by the heat of the flames, and tender seedlings attract bears. After the fire, life begins anew.

Animals pictured (left to right): black bears, least chipmunk, pronghorn antelope, downy woodpecker. Plants pictured: birch sapling.

 INTERMOUNTAIN GRASSLANDS

A plain of open land shimmers in grays and greens between the Rocky Mountains to the east and the Sierra Nevada and Cascade ranges to the west. Commonly called grasslands, this dry terrain is actually sagebrush **scrubland**, because more sagebrush grows here than grass. Sagebrush is a bush with narrow, silvery leaves.

Animals that can eat sagebrush leaves instead of grass make their homes here. Elk and pronghorn antelope feed on the shrubs, especially in winter, when they can find little else. Voles and ground squirrels burrow underground and create tunnels that protect them from predators such as coyotes.

Animals pictured (left to right): elk, killdeer, Jack rabbit, coyote, sagebrush voles.
Plants pictured (left to right): sagebrush and bunchgrass.

9 SONORAN DESERT

The sun scorches the desert sand in this dry, or **arid**, landscape, which extends from Mexico into southern California and western Arizona. Less than 10 inches of rain falls here each year. Daytime temperatures can soar to over 100 degrees Fahrenheit, then drop to nearly freezing after the sun sets.

Desert plants and animals must be able to survive the extreme temperatures and dryness. Cacti collect water during the brief rains and burst into bloom in the spring. In the mornings and late afternoons, cold-blooded reptiles warm themselves on the sand. But like other creatures, they seek shade during the hottest part of the day, saving energy for later, when the desert cools.

Animals pictured (left to right): horned lizards, Sonoran shovelnose snake, desert tortoise, Gila monster.
Plants pictured (left to right): fishhook cactus, beavertail cactus, claret cup cactus, cushion cactus.

Southwest of North America, the Hawaiian Islands rise from the Pacific Ocean. This is a **tropical** region close to the equator, an imaginary band that encircles the earth midway between the North Pole and the South Pole. Each island is surrounded by an underwater sandy ridge, or reef, covered with coral.

One-celled plants called algae live on the bodies of the coral animals and help to strengthen their skeletons. In return, coral lifts the plants toward the sun that filters through the water. It may take 500 years for a coral reef to grow just 7 feet. Other sea creatures, such as brightly colored tropical fish and sea snails, depend on coral for food and shelter from the wind, tides, and waves that shape the reef.

Animals pictured (left to right): finger coral, conch, reef triggerfish, blacktail wrasses, lined butterflyfish, venomous urchin, orange tube coral.

Cities may seem unlikely habitats for wild animals, but many creatures are at home among even the tallest buildings of the West. Urban animals do not depend on one source of food or shelter, but use everything they find around them. Because of this, they are called **generalists.**

For tree-dwelling animals such as squirrels and opossums, city parks seem like forests. Lakes and ponds, no matter how small, are wetlands that attract ducks and geese. And skyscrapers, like steep cliffs, provide roosts and nesting sites for pigeons, gulls, and even falcons. City wildlife thrives where other animals cannot.

Animals pictured (left to right): pigeons, common crow, gray squirrel, mallard ducks, Canada goose, Virginia opossum, raccoon.
Plants pictured: maple, oak, and linden trees.

Close your eyes for a moment.

Imagine what the West would be like
without the variety of habitats you have just visited.
Many of the habitats in the West are in trouble.
There are more people than ever before. As cities grow, wild lands shrink.
Plants and animals are in danger of losing their homes.

What can you do to help save the wild West?
First, you can learn more about a habitat near your home.
The best way to do this is by observing. Pay attention to the seasons
and the weather. Think about the landscape: Do you live in the mountains
or the desert, close to wetlands or the ocean? Do you have a garden or live
in a city with a park? What kinds of plants do you see there? Look up at the
trees and down at the soil, and watch for animals. You can use field guides
to help you identify the plants and animals around you. Then, share
what you have learned with other people. You can join a conservation
group at school or get involved with a campaign to save a habitat.
Finally, remember that we live with many different plants
and animals. In a way, our whole world is a habitat.
And every living creature has a place on Earth.

GLOSSARY

Alpine: The mountain terrain that is above the tree line, the highest point above which trees will not grow, is known as the alpine zone. Conditions at this height are severe, with intense sunlight, extremely hot and cold temperatures, strong winds, and poor, rocky soil. Only plants that grow close to the ground can survive on mountaintops.

Arid: Areas that are generally hot and extremely dry are called arid. Arid habitats may seem lifeless, especially during the middle of the day, when the sun is strongest. The animals that live here are active in the early morning and late evening, when temperatures are cooler.

Diversity: When a habitat provides animals with different sources of food and shelter, it is said to offer diversity. Because Yellowstone National Park offers diverse habitats, animals were able to survive the fires of 1988 by moving to other areas of the park, where they could find food and shelter.

Erosion: This refers to the breaking down and wearing away of the earth's surface by waves, wind, water, and ice. After many years, rock erodes into sand that can be found in different habitats, from the ocean to the desert, and even in garden soil.

Generalists: When an animal or a plant is called a generalist, it means that it can find food, water, and shelter in more than one kind of habitat. Unlike specialists, which need certain foods and places to live, generalists can adjust to find what they need. The American crow, a generalist, can live on fruit, meat, and even scraps of food found in garbage. Crows nest anywhere there are trees. The woodpecker, however, is a specialist, and depends on insects for food and old or dead trees for nesting holes.

Habitat: The area in which a plant or animal lives is called its habitat. Here, living creatures find the food and shelter they need to survive.

Rain forest: Most of the rain forests in the world are near the equator, in the tropical regions of Africa, Asia, Central America, and South America. A rain forest receives about 80 inches of rain in a year. The only rain forests that are not tropical are found in the Pacific Northwest. Temperatures there are cooler than in tropical forests, but the two habitats are similar in amount of rainfall.

Scrubland: Scrublands are large, flat areas covered by low trees or bushes. Sometimes called cold deserts, these regions receive little rain but do not get as hot as desert habitats. A wide variety of plants, from sagebrush to bunchgrass to wildflowers, provide food and shelter for animals.

Tidal mud flats: These are expanses of mud that are exposed and covered daily by the changing tides. Cord grass and eelgrass root in the mud. In the winter, these grasses die back and are recycled into the mud, which often smells like rotten eggs as the plants decompose.

Tropical: The regions of the earth that lie within 1,600 miles north or south of the equator are called the tropics. Temperatures in tropical areas are high most of the year for two reasons. First, the sun shines directly overhead at noon, warming the area more than slanted sun rays would. Second, the sun shines 12 hours a day, without the seasonal changes that exist farther from the equator.

Tundra: Tundra is the vast, flat land that lies at the northernmost points of Europe, Asia, and North America. The tundra is never completely warmed by the sun. Winters remain dark. The summer sun, which is visible all night, is weak this close to the North Pole because its rays hit the earth at a slant. Three or more feet below the surface of the tundra is permafrost, a layer of soil half a mile thick that is always frozen, even during the summer, when the upper crust of snow and ice melts.

Wetlands: These are areas that are wet for at least a few weeks every year. They range from deep lakes to shallow holes that may hold rain but are otherwise often dry. Wetlands are important to the survival of many plants and animals. They serve as filters, since wetland plants help remove impurities from water that seeps into the ground or flows into streams, rivers, and even the ocean. Wetlands also act as sponges, soaking up extra water in their porous soil, which helps prevent floods.